ADDRESS

OF

PROFESSOR J. S. NEWBERRY,

PRESIDENT OF THE AMERICAN ASSOCIATION FOR THE YEAR 1867,

ON RETIRING FROM THE DUTIES OF PRESIDENT.

Gentlemen of the American Association for the Advancement of Science: —

Every day of our lives we hear that this is an age of *progress;* and that it is so we find evidence at every turn. The rapidity with which effects follow causes in human events, the celerity with which the plan is carried into execution, gives to a *year*, in the experience of one of the present generation, the practical value of a lifetime in ages past. Much labor has been expended on the exposition of the causes of the mental activity of the present age, and of the grand achievements which have attended it; and yet the key to the whole enigma is to be found in the universal adoption of the comparatively new system of inductive reasoning. It would be foreign to my purpose to attempt to illustrate or defend this proposition, and I must, therefore, trust to its acceptance without argument, while we pass to consider that branch of the subject which more immediately demands our attention.

Although the progress of the age to which I have referred has been a matter of wonder and delight to all students of humanity and civilization, many of our best men have been somewhat alarmed and dizzied by it; and, while accepting the achievements of modern industry and thought as full of present good and future promise, they are not a little concerned lest our railroad speed of progress should lead to its legitimate consequences, a final crash — not of things material, but of those of infinitely more value — of opinions and of faith. As often as it is boasted that this is pre-eminently an age of progress, that boast is met by the inevitable "but" (which qualifies our praise of all things earthly) "it is equally an age of scepticism." For the truth of this assertion the proof is nearly as palpable as of the other; and in view of the ruthlessness with which the man of the present removes ancient landmarks and profanes shrines hallowed by the faith of centuries, it is not surprising that many of the good and wise among us should deplore a liberty of thought leading, in their view, inevitably to license, and mourn over this wide-spread scepticism as an evil and inscrutable disease that has fallen upon the minds and hearts of men.

Now for every consequence there must be an adequate cause; and while confessing the fact of this modern lack of faith, I have thought that a few moments given to an analysis of it, and an attempt to trace it to its source, might not be wholly misspent, — might possibly, indeed, result in giving a grain of encouragement to those who look with distrust and dread upon the investigations and discussions which now occupy so large a portion of the time and thought of our men of science.

If the wheels of time could, for our benefit, be rolled back, and we could see in all its details the civilization of Europe three or four hundred years ago, we should find that our so much respected ancestors, who fill so large a space on the page of history, were little better than barbarians. Among the English, the French, the Germans, Spanish, and Italians we should find a phase of civilization which, excepting that it included the elements — as yet but imperfectly developed — of a true religious faith, is scarcely to be preferred to that of the Chinese. Aside from the vast difference perceptible between the civilization of that epoch and ours, as exhibited in the political condition of the people, in their social economy and morals, the general intellectual darkness of the period referred to could not fail to impress us both profoundly and painfully. Out of that darkness and chaos have come, as if by magic, all our modern democracy, with its individual liberty and dignity, all our civil and religious freedom, all our philanthropy and benevolence, all our diffused comfort and luxury, most of our good manners and good morals, and all the splendid achievements of our modern scientific investigation.

It is unnecessary for me here to describe in detail the origin and growth of modern science. That has been so well done by Dr. Whewell, that all men of education are familiar with the steps by which the grand, beautiful, and symmetrical fabric formed by the grouping of the natural sciences has acquired its present lofty proportions.

Previous to the period when the Baconian philosophy was accepted as a guide in scientific investigation, but one department of science had attained a development which has any considerable claim to our respect. Mathematics, both pure and applied, had been assiduously cultivated from the remotest antiquity, and with a degree of success which has left to modern investigators little more than the elaboration of the thoughts of their predecessors. In Metaphysics, — which had claimed even a larger share of the attention of the scholars of antiquity, — little progress had been made. Perhaps I am justified in saying little progress was possible, inasmuch as in the light of all the great material discoveries of modern times the metaphysicians of the present day are debating, with as little harmony of opinion, the same questions that divided the rival schools of the Greeks. Each successive generation has had its two parties of idealists and realists, who have discussed the intangible problems which absorbed the great minds of Plato and Aristotle with a degree of enthusiasm and energy — and it may be of acrimony — which seems hardly compensated by any expansion of the human intellect or amelioration of the condition of mankind.

Of the *physical sciences* we may say that, except Astronomy, no one had an existence prior to the time of Bacon. There were men of vast learning, and much that was *called* science in the mass of reported observation that had been accumulating from century to century, until it had become "*rudis indigestaque moles;*" but in this — though it constituted the pride of universities, the intellectual capital with which the savant thought himself rich, and that on which the professional man depended for success — there was far more error than truth, and its study was sure to mislead and likely to injure. In these circumstances the task before the scientific reformer was one more difficult than that of clearing the Augean stables; no less, in fact, than to seat himself before this great heap of rubbish, this mass of truth and error, — of the sublimest philosophy with the wildest fiction, — to patiently winnow out the grains of truth, and from infinitesimal facts build up a fabric that should have a sure foundation below, and beauty and symmetry above. What more natural, then, than that the process adopted in winnowing this chaff-heap should be that which had given success to the only true science of the period? — that the mathematical touchstone should be the test by which every grain was tried? And such precisely was the course pursued; perhaps we may even say the only one practicable. Provided with this test, the reformer was compelled to rejudge upon its merits every proposition submitted to him, and accepted only as true such as could be demonstrated. The materials which composed the science to be reformed naturally fell into several categories. First, — That which had been *demonstrated* to be true. Second, — That which was *demonstrable.* Third, — That which was *probable.* Fourth, — That which was *possible,* and Fifth, — That which was *impossible.* Of these he systematically rejected all but the first and second classes. And this, in few words, has been the method adopted, not only in the purification of old science, but in the creation of new.

It will be seen at a glance, that in this process, all that was contrary to the order of nature (supernatural or spiritual) was necessarily excluded; and it was taken for granted that the mathematical or logical faculty of the human mind was capable of solving *all* the problems of the material universe. Sir William Hamilton and others have demonstrated the inadequacy of mathematical processes as a guide to human reason, and a moment's thought will show us that our boasted intellect is incapable of grasping even all the material truths which are plainly presented to it. To illustrate: as we scan the heavens of a clear evening, we recognize the fact that we stand, as it were, on a point in space, where our field of vision is limitless; the heavenly bodies stretching away into the realms of obscurity, and becoming invisible only through the imperfection of our organs of vision. Bringing to our aid the most powerful telescopes, we are apparently as far as ever from reaching the limits of the universe; and when we endeavor to *conceive* of such a limit, the reasoning faculty finds itself incapable of grasping either of the two alternatives offered to it, one or the other of which must be true. The universe must be either limited or limitless. But no man can conceive of a universe without a limit; and if it be regarded as terminated by definite boundaries, the imagination strives in vain to fill the void which reaches beyond. In fact, we stand here face to face with infinity, and recognize the fact that the infinite exists without the power to comprehend it.

The same is true of time. We cannot conceive of its beginning or its end. All things which come within the scope of our senses are limited in duration and circumscribed in space; and though we prate flippantly of the infinite, the pretence that we can grasp it is simply idle talk.

Conducted on such a plan, it was inevitable that scientific investigations should be narrow and materialistic in their tendency. No matter how strong the probability in favor of the truth of a certain proposition, — though the whole fabric of society were based upon its acceptance, and it formed the foundation of civil and moral laws; even though it controlled the actions of the philosopher himself, — if not proved consistent with nature's physical and material laws it must be rejected as unworthy to enter into the construction of the edifice he was erecting. In his great task of undoing the work of blind, unreasoning faith, and wild, illogical speculation, all the fruit of such faith or speculation must be looked upon as matter valueless to his hand. We may even go further and say that were it true that the Supreme Intelligence *had* created the material universe, and by special providence modified or thwarted the general laws through which that universe was governed, such divine supervision and such miraculous interposition must necessarily have been ignored.

Let it not be inferred, however, that each and all of the great men who have been engaged in this work of scientific reformation were necessarily driven to be impious iconoclasts, or that, in their efforts to emancipate themselves from time-honored errors, they necessarily prostituted the liberty they gained to selfish or sensual purposes. On the contrary, the most important advances which the human intellect has made within these later centuries have been due to the efforts of men of the purest and most conscientious character, — men whose lives were devoted with the utmost singleness of purpose to determine *what is truth*, — men who, knowing that all truth must be consistent with all other truth, were willing to go whithersoever it should lead. If it shall prove that they have been occupied with "mint, anise, and cumin," omitting the "weightier matters of the law," it is also true that in no other way could the material laws of the universe be thoroughly investigated than by making them the subjects of an absorbed and undivided attention. It would be as just to impugn the motives and decry the merits of the maker of our almanacs, because his mathematical calculations were not interlarded with moral maxims, as to reproach the student of natural phenomena because he did his work so well, and left to others the co-ordination of the results of his efforts with the accepted dogmas of religious faith.

And it is not true, in any sense, that these devotees of science have lived in vain; for to them we mainly owe the fact that man is not only wiser now than formerly, but that he is better and happier.

In justice to the man of science we must claim for him the position of co-laborer with, and indispensable ally to, the philanthropists and moralists; for from no source have they drawn richer lessons, stronger arguments, or more eloquent illustrations than from his discoveries.

And yet, while conceding conscientiousness and purity of motive to the vast majority of our men of science, and acknowledging the contributions they have made, and are making to human happiness; compelled by my sense of justice to defend their spirit, approve their methods, admire their devotion, and assert their usefulness, I cannot deny that the tendency of modern investigation is decidedly materialistic. All natural phenomena being ascribed to material and tangible causes, the search for and analysis of these causes have begotten a restless activity and an indomitable energy which will leave no stone unturned for the attainment of their object. But while this is apparent, and, indeed, inevitable, as has been seen from the sketch of the growth of modern science, I am far from sharing the alarm which it excites in the minds of many good men. Nor would I encourage or excuse that spirit of conservatism — to call it by no harsher term — which, for the safety of a popular creed, would, by any and all means, repress, and, if possible, arrest investigations that may, it is feared, become revolutionary and dangerous.

Such opposition, in the first place, must be fruitless. All history has proved that persecution by physical coercion or obloquy is powerless to arrest the progress of ideas, or quench the enthusiasm of the devotees of a cause approved by their moral sense. The problems before our men of science *must* be solved in the manner proposed, if human wisdom will suffice for the task. In every department of science are men actuated simply by a thirst for truth, whom neither heat nor cold, privation nor opposition, will hold back from their self-appointed tasks. We may, therefore, accept it as a finality, that this problem will be carried to its logical conclusion.

In the second place, if possible, the arrest of scientific investigation would be not only undesirable, but an infinite calamity to our race. As has been so often said, truth is consistent with itself. If, therefore, our faith in this or that is based on truth, we have no cause for fear that this truth will be proved untrue by other truths. And more than this: it seems to me that, in the reach and thoroughness of this material investigation, we may hope for such demonstration of the reality of things immaterial as shall produce a deeper and more universal *faith* than has ever yet prevailed.

Through this very spirit of scepticism which pervades the modern sciences we are compelled to exhaust all material means before we can have recourse to the supernatural. When, however, that is done, and men have tried patiently and laboriously, but in vain, to refer all natural phenomena to material causes, *then,* having proved a negative, they will be compelled to accept the existence of truth not reached by their touchstone, and faith be recognized as the highest and best knowledge.

That such will be the result is the confident expectation of many of the wisest of the scientific men whose influence is looked upon with such alarm by those who, in their anxiety for their faith, demonstrate its weakness.

Already, as it seems to me, scientists have reached the wall of adamant — the *inscrutable* — that surrounds them on every side, and, ere long, we may expect to see them return to that heap of chaff from which the germs of modern science were winnowed, with the conviction that there were left buried there other germs of other and higher truths than those they gleaned, — truths without which human knowledge must be a dwarfed and deformed thing.

A few illustrations from the many that might be cited will suffice to show the materialistic tendency of modern science. In "Pure Philosophy," — as the students of Psychology are fond of styling their science, — the names alone of Comte, Buckle, Herbert Spencer, Mill, and Draper will suggest the more prominent characters of the school they may be said to represent. The most conspicuous feature in the "Positive Philosophy" of Comte is the effort it exhibits to co-ordinate the laws of mind with those of matter. Spencer is a thorough-going mental Darwinist, who considers the highest attributes of the human mind, the loftiest aspirations of the soul, as only developed instincts, as these were but developed sensations. Mill, more guarded, more fully inspired with the spirit of the age, — which *believes* nothing, and is a foe to speculation, — leaves the *history* of our faculties to be written, if at all, by others; takes them as they are, but reasons of conscience and free-will with an independence of popular belief that savors more of the material than the spiritual school. Buckle wore himself out in a vain chase after an *ignis fatuus,* an inherent, inflexible law of human progress, and hence of human history. Draper is a developmentist, but not a Darwinian. With him civilization is a definite stage in the growth of mind; a degree of development to which it is impelled by a *vis a tergo*, not unlike, in kind, to that which evolves from the germ, the bud, the leaf, the flower, and the fruit in plant-life, — a development which, when unchecked and free, will be regular and inevitable, but which is so modified by the accidents of race, climate, soil, geographical position, etc., as to render it difficult to say whether the rule or the exception has, in his judgment, greatest potency. If he were a consistent Darwinist, the accidents of development would be its law.

Among the students of "Social Science," — a new and important member of the sisterhoood of sciences, — as in most of the other departments of modern investigation, two groups of devotees are found; one patiently and conscientiously studying the problems of social organization, inspired with the true spirit of the Baconian Philosophy, ready to follow whithersoever the facts shall lead, and having for their object that noblest of all objects, the increase of human happiness. The other class of investigators, in whom the bump of destructiveness is largely developed, would be delighted to tear down the whole fabric of society, and abrogate all laws, both human and divine. Looking upon man as literally the creature of circumstances, as an inert atom driven about by material forces, conscience and responsibility are by them repudiated, and laws and penalties regarded simply as relics of barbaric despotism. The dreary soul-killing creed of these fatalists is fortunately so repugnant to the reason and feelings of the majority of men, that there is little danger that their efforts will reach their legitimate conclusion in throwing society into a state of anarchy and chaos.

In Theology or Biblical Science, the tendency of modern investigation is so distinctly felt, that I need only refer to it. The spirit of independent criticism, so noticeable elsewhere, is still more conspicuous here; assuming sometimes the form of derisive scepticism, but oftener of cold, passionless judgment on the reported facts of sacred history, or the psychological phenomena of religious faith, studied simply as scientific problems.

The names of Strauss, Renan, and Colenso will suggest the results to which men, possibly honest, are led by this so-styled "enlightened and emancipated spirit of inquiry;" while "Ecce Homo" and cognate productions may be considered as the fruit of this spirit, tempered by a very liberal but apparently sincere faith.

Aside from these more marked examples of the decided "set" in the tide of modern religious opinions, we everywhere see evidences that no part of the religious world is unmoved by it. In every sect and section an impulse is felt to substitute for abstract faith, the "faith without works," — rather a characteristic of the religion of our fathers, and not unknown at present, — that other faith which is evidenced *by* works. In other words: in our day more and more value is being attached to *this life*, as a sphere for religious effort and experience. With what propriety, I leave to the individual judgment of my auditors; the faith of every sect and man is coming to be respected and valued precisely in the ratio of the purity, unselfishness, and active sympathy in the life produced by it.

While, therefore, we have less now than formerly of the self-centred and fruitless piety of the old deacon, who excused his avarice by proclaiming that "business was one thing and religion another, and he never allowed them to interfere," in place of that we have many an Abou Ben Adhem, and all the splendid exhibitions of modern philanthropy.

Though the golden mean displayed in the life and words of Christ is far better than either extreme, I cannot but think the present religious condition of the world is better than any which has preceded it.

In Ethnology, — the pre-historic history of the human race, — the researches of a large number of investigators who are devoted to its study have made interesting and important additions to our knowledge; but it cannot be denied that the result of such investigation has been to create general distrust of our previously accepted chronology, and give an antiquity to man such as the scholars of a previous generation would have looked upon as not only unwarranted but impious. It should be said, however, that our preconceived opinions of the antiquity of the human race — like those of the age of the earth itself — were based upon no solid foundation in nature, history, or revelation; and that our system of chronology was a matter of convention, about which there has been a wide latitude of opinion among the scholars of all ages.

In regard to the *origin* of man, — whether by special creation or by development, — we may confidently assert that modern investigation has given us no new light. Among those who have accepted the theory of a special creation, and have differed only in regard to the number of species and their places of origin or centres of creation, there has been such a diversity of opinion that all confidence in the reality and value of the bases of their reasoning has been lost.

Among the advocates of a multiplicity of species and diversity of origin we have from Blumenbach to Agassiz almost every number between fifteen and three as that of distinct species of the human race, scarcely any two writers advocating the same number. We may, therefore, very fairly infer that the facts upon which their conclusions are founded are not of a very clear and unmistakable character.

The subject of the origin of the human race brings us into the domain of zoölogy, and opens the wide question of the origin of species, which, of late years, has been shaking the moral and intellectual world as by an earthquake. While the various writers upon the origin of the human race were gathering with so much industry, and reporting with so much eloquence the proofs of a diversity of origin, the Darwinian hypothesis comes in and refers, not only all the human family, but all classes of animals and plants, to an initial point in a nucleated cell.

It would be impossible for any one to discuss in a fair and intelligent manner the great question of the origin of species, in anything less than a bulky volume. The merest mention is, therefore, all we can give to it at the present time. Although the appearance of Darwin's book on the Origin of Species communicated a distinct shock to the prevalent creeds, both religious and scientific, the hypothesis which it suggests, though now for the first time distinctly formularized, was by no means new; as it enters largely into the less clearly stated development theories of Oken, Lamarck, De Maillet, and the author of the "Vestiges of Creation." There was this difference, however, that in the developmental theories of the older writers the element of *evolution* had a place; the process of development had its main-spring in an inherent growth, or tendency, such as produces the evolution of the successive parts in plant-life, while, according to Darwin, the beautiful symmetry and adaptation which we see in nature is simply the form assumed by plastic matter in the mould of external circumstances.

Although this Darwinian hypothesis is looked upon by many as striking at the root of all vital faith, and is the *béte noire* of all those who deplore and condemn the materialistic tendency of modern science, still the purity of life of the author of the "Origin of Species," his enthusiastic devotion to the study of truth, the industry and acumen which have marked his researches, the candor and caution with which his suggestions have been made, all combine to render the obloquy and scorn with which they have been received in many quarters peculiarly unjust and in bad taste. It should also be said of Mr. Darwin that his views on the origin of species are not inconsistent with his own acceptance of the doctrine of Revelation; and that many of our best men of science look upon his theory as not incompatible with the religious faith which is the guide of their lives, and their hope for the future. To these men it seems presumption that any mere man should restrict the Deity in His manner of vitalizing and beautifying the earth. To them it is a proof of higher wisdom and greater power in the Creator that He should endow the vital principle with such potency that, pervaded by it, all the economy of nature, in both the animal and vegetable worlds, should be so nicely self-adjusting that, like a perfect machine from the hands of a master maker, it requires no constant tinkering to preserve the constancy and regularity of its movements.

This much I have said in view of the possible acceptance of the Darwinian theory by the scientific world. I should have stated *in limine*, however, that the Darwinian hypothesis is not, and can never be fully accepted by the student of science who is inspired with the spirit of the age. From the nature of things it can be proved only to a certain point, and while we accept that which is, or can be, proven, — and for it sincerely thank Mr. Darwin, — that which is hypothesis, or based only upon probabilities, we reject, as belonging in the category of mere theories, to disprove or purify which the modern scientific reform was inaugurated. Much, too, may be said against the sufficiency of "natural selection in the struggle of life," from observations made upon the phenomena of the economy of nature. Necessarily, the action of the Darwinian principle must be limited to the individual, be literally and purely selfish; and if it can be proved that a broader influence pervades the created world, that something akin to benevolence enters into the organization of the individual, — something which benefits others and not himself, — one single fact establishing this truth would hurl the entire Darwinian fabric to the ground, or rather restrict it to its proper bearing upon the limits of variation, and the mooted question of "what is a species?"

One of the most potent influences in the perpetuation of species is fecundity in the individual, whereas we see in social insects the economy of the community is best served by a total loss of this power in the great majority of the individuals which compose it. This objection will perhaps be met by the Darwinians with the assertion that the community, in fact, constitutes an individual; but I must confess that I find it difficult to comprehend how the sterility of the workers in ants and bees was ever introduced through the medium of modified descent, the Darwinian method, or how it is kept up from generation to generation among those individuals which have no posterity to inherit their peculiarities of structure.

The Honey Ants of Mexico offer additional difficulties. Among them a portion of the community secrete honey in the abdominal cavity until they resemble small grapes, and these individuals, during the winter, are despatched in succession to furnish food for the other members of the colony. How, by modified descent, is this honey-making faculty transmitted, when those who possess it are systematically destroyed?

A still harder nut for the Darwinians to crack is furnished in a fact stated by Dr. Stimpson, that among the crustacea, which do not live in communities, a very large proportion of the individuals of a numerically powerful species pass their lives as neuters or undeveloped females.

Another element in nature's economy, which at first sight suggests an objection to the Darwinian theory, is that of *beauty*, which affects others far more than the possessor. This is considered by the Darwinians simply as an attraction to the opposite sex, but as a fact we find that in the larval condition of some insects — a condition in which no propagation is effected — varieties of form and combinations of color exist which appeal to our sense of beauty scarcely less forcibly than in the perfect insects.

Again, the origin of life is left completely untouched by the Darwinian hypothesis, and so long as the vital principle resists, as it has done, all the efforts of theorists and experimenters to bring it within the category of material forces, so long we must regard the world of life as including elements not amenable to the laws which control simple inert matter.

Upon this question of the origin of life so much is being done and said that you will expect a word of reference to it at my hands, yet little more can be reported as the result of modern research than that the origin of life is as great a mystery as ever. You will all remember how, a few years since, we were startled by the announcement of the discovery of the generation of the "*Acarus Crossii;*" and, while our original distrust of the accuracy of the observations of Mr. Crosse was strengthened by the failure of subsequent experimenters to reproduce his results, our unbelief is further confirmed by the unanimity of all the more modern and intelligent devotees of spontaneous generation in the assertion that life can only originate in its simplest form, that of a unicellular organism. There is no Darwinist who will concede the possibility of an animal as highly organized as an *Acarus*, with body, head, limbs, digestion, and senses, all more or less complete, being the product of spontaneous generation and not the result of slow and gradual development.

Still farther; it is known that the animal kingdom rests upon the vegetable as a base. Animals being incapable of assimilating inorganic matter could not exist without plants. Plants must therefore have preceded animals, and the fruit of spontaneous generation must be a protophyte and not a protozoan.

Yet, notwithstanding its difficulties, there are to-day men, respectable by their numbers and attainments, who are believers in spontaneous generation; but it is with this proviso, — which leaves the mystery as great as ever, — that only from *organic* matter can organisms be produced. So that to the original and primary appearance of life upon the earth, modern science has given us not the slightest clew.

As I have said, the materialists have so far utterly failed to coordinate the vital force with those which we designate as material. The beautiful and important discoveries which have followed researches into the correlation and conservation of forces by pointing to a unity of all the forces in the material world have naturally prompted efforts to centralize, with electricity, magnetism, and chemical affinity, that which we know as vital force. But a moment's reflection will show us how far removed is this vital force from all others with which it has been compared.

The nicest manipulations of chemical science will probably fail to detect a difference in composition between the microscopic germs of two cryptogamous plants. Each consists of the same elements, — carbon, nitrogen, hydrogen, and oxygen, in nearly or quite the same proportions. Both may be planted in a soil which laborious mixture has rendered homogeneous, and subsequently supplied with the same pabulum, and yet, in virtue of some inscrutable, inherent principle, one develops an humble moss, and the other rises into the beauty, symmetry, and even grandeur of a tree fern. The same lesson is taught by the spermatozoa of the mouse and the elephant. Indeed, all the phenomena which attend the reproduction of species are totally at variance and incompatible with those which mark the action of material laws. Why, in physical circumstances differing *toto cœlo*, does the germ produce a plant or animal so closely copying the parent? and whence this tenacity of purpose in the germ which reproduces, through a long line of posterity, the trivial characteristics of a remote ancestor? Even within our limited observation, we have been struck by the reappearance in the grandchild of the voice, the gesture, the stature, the features, or some other marked peculiarity of his grandsire. Whence comes the force of the axiom that "blood will tell"? — and how incomprehensible that, by the action of only material laws, mental force, or, it may be, moral infirmity, is transmitted from generation to generation, in spite of the system of infinitesimal dilution through which it passes!

And now, even with this hurried and sadly imperfect exposition of the tendency of modern science, the time at our command has been consumed. Before leaving the subject, however, I crave your indulgence for a word to those who, wholly absorbed in the study of the laws which regulate the material universe, are so deeply impressed with their universality and potency that they forget that law is but another name for an order of sequence, and has in itself no force. These are they who, in their pride in the achievements of the human intellect, fail to realize that the universe furnishes conclusive proof that all our philosophy, all our logic, all our observation, are utterly inadequate to solve the problems that are presented to us, — inadequate not simply from the limited nature of our powers of observation, but because the human mind, though forced to confess the existence of the infinite, is utterly unable to grasp it; and that while the logic of reason and the logic of numbers suffice for a qualified understanding of the manner in which material forces work, of the origin and nature of these forces we are and must ever remain ignorant, unless gifted with higher powers than we now possess. As has been stated, seen from the stand-point of our modern materialists, and judged by the criteria which they have adopted, spiritual existence and supernatural phenomena, even if as all-pervading as the most devout religionist believes, must, from *a priori* considerations, be utterly ignored. Of those who are thus led by their regard for the dignity of material laws to reject the idea of a creative and overruling Deity, I would ask, Is not man himself a disturbing element in your universe? Whatever may be said in regard to man's free-agency, and however confidently it may be asserted that his will is but the resultant of the various motives that operate as distinct forces upon it, consciousness lies at the basis of all reasoning; and the conduct of every man proves that he accepts this axiom. As he issues from his door, he is conscious, beyond all argument, that it is in his power to turn to the right or to the left; and while he holds himself responsible for his volition, he cannot blame us if we ascribe to him free-agency. Man is, therefore, an independent power in the universe. He wills and creates. The locomotive is as truly his creation as himself, fashioned from the dust of the earth and vitalized by the breath of the Almighty, is the work of His hands. If, therefore, all the realm of nature is controlled through material laws, by forces that, like attraction, electricity, chemical affinity, etc., act in an invariable and inflexible way, in this universe man is a stupendous anomaly; and unless he can be degraded from his position of pre-eminence in this material world, the boldest and most irreverent of modern philosophers will strive in vain to dethrone the Great Creator from the rule of the universe, or from His place in the hearts and minds of men.

In consequence of the absence of President F. A. P. Barnard in Europe, an arrangement was made according to which Professor Newberry delivered the address at Burlington, and President Barnard is expected to give his own at the next meeting in Chicago, Illinois.

www.ingramcontent.com/pod-product-compliance
Lightning Source LLC
LaVergne TN
LVHW020641110826
845149LV00004B/1307

* 9 7 8 1 4 1 8 1 9 0 5 4 5 *